THE SEASONS IN MR GREEN'S GARDEN

by Ruth Owen

RUbY TUesdaY BOOKS

Published in 2024 by Ruby Tuesday Books Ltd.

Copyright © 2024 Ruby Tuesday Books Ltd.

All rights reserved. No part of this publication may be reproduced in whole or in part, stored in any retrieval system, or transmitted in any form or by any means, electronic, mechanical, photocopying, recording, or otherwise, without written permission from the publisher.

Editor: Mark J. Sachner
Design: Emma Randall
Production: John Lingham

Photo credits:
Alamy: 14BR (imageBroker.com), 19T (Chris Mattison); Emma Bowring: 11T; Dwight Kuhn Photography: 11C; Nature Picture Library: 10B (Nick Upton), 13T (Stephen Dalton), 16B (Stephen Dalton), 17T (David Tipling); Shutterstock: Cover (PJ Photography/Peter Baxter/Valerie Quemener/Alexander Raths), 5 (x.marynka/Pakhnyushchy/Wirestock Creators/Leonid Ikan), 6 (Sternstunden), 7 (Olena Belevantseva/Emilio100/sevenke), 9 (Kaczor58/Madlen), 10 (NCAimages/thatmacroguy), 11B (Eric Isselee), 12 (Zanna Pesnina/mehmetkrc), 13 (Maciej Olszewski/Vita Serendipity/Holger Kirk), 14BL (Richard Griffin), 15 (krigo), 16 (Erik Steinebach/Brano Molnar), 17C (Clare Lusher), 18 (Bachkova Natalia/Michael Schroeder); 19C (PJ Photography/Stphanie CROCQ), 20B (FotoDuets); Superstock: 17B (Sumio Harada, Minden Pictures), 19B (Loop Images).

British Library Cataloguing in Publication Data (CIP) is available for this title.

ISBN 978-1-78856-346-8

Printed in Poland by L&C Printing Group

www.rubytuesdaybooks.com

Contents

Welcome to the Garden 4

How Do We Know It's Spring? 6

What Is Happening Underground? 8

The Bumblebees Are Back! 10

It's Time to Wake Up and Eat! 12

What Does the Oak Tree Do in Spring? 14

Eggs, Hunting and New Babies 16

Why Are the Garden Birds So Busy? 18

Sunshine, Seeds and Plans for Summer 20

Mr Green's Garden Glossary 22

Index 24

In a garden, there is always something to look forward to. . . .

Welcome to the Garden

Hello!

Hello!

This is Mr Green and his friend and neighbour, Mo.

Together, they cleaned up an ugly, rubbish-filled place in their neighbourhood.

THE SHED

They turned this place into a garden for all their neighbours to enjoy.

Mo says:

A garden is a very good place to learn all about **seasons**, and watch them change.

Every year there are **4** seasons.

Each one lasts for about three months.

Spring...
March April May

Summer...
June July August

Autumn...
September October November

Winter...
December January February

Mr Green says:
Let's discover what happens in the garden in spring!

How Do We Know It's Spring?

Each day during spring, there is a little more daytime and less night-time.

As winter ends and spring begins, the weather gets warmer.

In daytime it can be sunny. However, at night, it might still be cold.

Mo says:
In spring, there is often lots of rain.

Spring Weather

In spring, plants that have been resting all winter start to grow again.

Green **buds** sprout from bare plant **stems** and tree **branches**.

Stem

Bud

Unfolding leaves

Buds unfold and become new leaves or flowers.

Bud

Cherry tree

Cherry tree flower

Mr Green says:
Grass starts to grow in spring. We mow some grass and leave the rest long as a home for **insects** and other small animals.

What Is Happening Underground?

Some spring plants, such as daffodils, grow from underground parts called **bulbs**.

In spring, green **shoots** grow from the bulbs and push up into the sunlight.

The shoots become leaves and stems with flowers.

Hyacinth · Tulip · Tulip bud · Daffodil · Crocus · Stem · Leaf · Roots · Bulb

A bulb contains food that gives the plant energy to grow.

Under the soil is an acorn that was buried by a squirrel in autumn. Inside the acorn there is an oak tree **seed**.

In spring, tiny **roots** and a shoot sprout from the seed.

Shoot
Acorn
1

Shoot
Leaf
2

Oak tree seedling
Stem
Acorn
Roots
3

Mr Green says:
An acorn shoot becomes a little **seedling** that will grow into a new oak tree.

The Bumblebees Are Back!

All winter, queen bumblebees have been **hibernating** in the garden. Now, each queen crawls from her hiding place.

A queen bumblebee visits lots of spring flowers to drink **nectar** and collect **pollen**.

Pollen

Bumblebee

Crocuses

She finds a place to be her nest, such as an empty mouse hole or in a compost heap.

Nest hole

It's Time to Wake Up and Eat!

As the weather warms up, snails slide from their winter hiding places. They munch on new shoots and leaves.

Snail

The garden's ladybirds have been dormant all winter.

Ladybird

Aphid

Being dormant is a little like being in a deep sleep.

Now the hungry beetles hunt and eat tiny insects called aphids.

A cabbage white butterfly has been in the pupa stage of its life cycle. Now the pupa changes into an adult.

Pupa

Pupa case

Butterfly

On a warm day, the new butterfly climbs from its pupa case.

Long mouthpart for drinking

The butterfly flutters from flower to flower drinking nectar.

Eggs

In spring, cabbage white butterflies meet up and **mate**.

Then each female lays hundreds of tiny eggs on leaves.

Butterfly laying eggs

What Does the Oak Tree Do in Spring?

In winter, an oak tree drops its leaves and takes a long rest.

In spring, new leaves sprout from its bare branches.

The tree also grows male and female flowers.

Mr Green says:

The male oak tree flowers release dusty pollen into the air.

The pollen floats on the breeze and lands on the tiny female flowers.

Then the female flowers are ready to produce new acorns.

New leaf

Male flowers

Female flower

Eggs, Hunting and New Babies

In spring, male and female frogs meet up and mate in the garden pond.

Each female frog lays up to 2000 eggs.

Egg

Frog

Each tiny black egg is inside a round blob of soft, clear jelly.

Jelly

After a few weeks, an egg becomes a tadpole that wriggles out of the jelly.

Tadpole

A hedgehog has been hibernating in the garden. Now she is awake and goes hunting at night.

She eats worms, slugs, beetles and centipedes.

Hedgehog

Female squirrel
Nesting material

Squirrel kit

Mr Green says:

Two squirrels met and mated in the garden. Then the female squirrel chose a hole in a tree to be her safe, cosy **den**.

Inside the den, she gave birth to four blind, hairless squirrel kits.

Why Are the Garden Birds So Busy?

In spring, it's the season for birds to build nests, lay eggs and raise chicks.

Father sparrow feeding a chick

A pair of sparrows has filled a nest box with dry grass, hair, string and feathers. Sometimes, sparrows pluck feathers from live pigeons for their nest!

Mr Green says:

Many chicks eat insects, worms and other minibeasts to help them grow.

Birds raise their chicks in spring or summer because there are lots of minibeasts around.

A female robin builds a cup-shaped nest from moss, grass, animal hair and dead leaves.

Moss

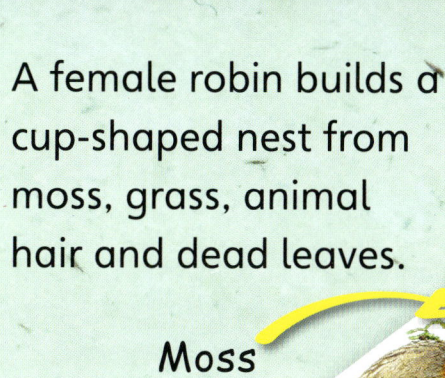

Female robin

She lays up to six eggs in the nest and sits on them to keep them warm. After two weeks, chicks hatch from the eggs.

Robin eggs

When the robin chicks hatch, their parents bring them insects to eat.

Sunshine, Seeds and Plans for Summer

Under the soil, seeds have been waiting all winter. They dropped from flowers last summer and autumn.

As the weather warms up, tiny shoots sprout from the seeds.

Shoots

Wildflower seeds

Mr Green says:
Spring is also the time of year when gardeners dig the soil and plant seeds.

Bean seeds

Mr Green's Garden Glossary

branch
A long stem that grows from the trunk, or main stem, of a tree.

bud
A new part of a plant that grows into a shoot, leaf or flower.

bulb
A round underground part that some plants grow from. Food for the new plant is stored in the bulb.

den
An animal's home. Female animals may give birth to their babies in a den.

hibernate
To spend the winter in a deep sleep without eating or drinking.

insect
A tiny animal with six legs and a body in three main parts. Most insects have wings.

larva
A fat, white young insect.

mate
To come together to produce young.

nectar
A sweet liquid made by flowers.

pollen
A coloured dust that is made by flowers, and is needed for making seeds.

pupa
One of the stages in the life cycle of some insects. For example, the four stages in a butterfly's life cycle are egg, caterpillar, pupa and adult butterfly.

roots
Underground parts of a plant that take in water from the soil.

season
One of the four parts of a year. The seasons are winter, spring, summer and autumn.

seed
A tiny part of a plant that contains all the material needed to grow a new plant.

seedling
A small new plant that grows from a seed.

shoot
A new part of a plant. Shoots grow from seeds, bulbs and the stems of plants.

stem
A long, thin part of a plant. Leaves and flowers grow from stems.

Index

A
animals 7, 9, 10–11, 12–13, 16–17, 18–19

B
birds 18–19
bulbs 8
bumblebees 10–11
butterflies 13

D
daytime and night-time 6, 17

E
eggs 11, 13, 16, 18–19

F
frogs 16

H
hedgehogs 17

I
insects 7, 10–11, 12–13, 17, 18–19

L
ladybirds 12

N
nests 10–11, 17, 18–19

P
plants 7, 8–9, 10, 12–13, 14–15, 17, 18–19, 20–21

S
seasons 5, 18
seeds 9, 14–15, 20–21
snails 12
squirrels 9, 15, 17

T
trees 7, 9, 14–15, 17

W
weather 6, 12–13, 14–15, 20–21

If you enjoyed this book, discover the story of how Mr Green and his friend Mo created their wonderful garden.

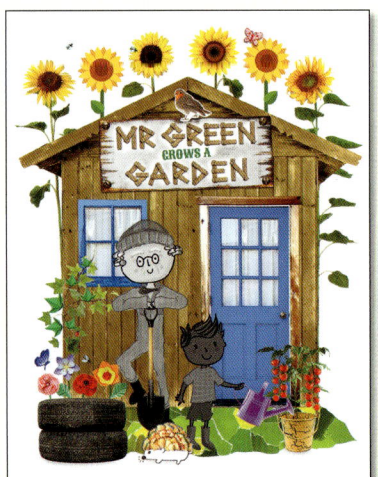

ISBN 978-1-78856-166-2